practise

Multiplication and Division

9-10

YEARS

Key Stage 2

Multiplication tables

Remember, multiplication can be done in any order.
This halves the number facts to remember! → $5 \times 7 = 7 \times 5$

Answer these as quickly as you can.

5 × 4 =	9 × 7 =	2 × 8 =	4 × 4 =
6 × 7 =	5 × 5 =	7 × 9 =	3 × 8 =
9 × 5 =	7 × 6 =	9 × 4 =	9 × 6 =
8 × 6 =	8 × 8 =	10 × 6 =	7 × 7 =
7 × 8 =	10 × 3 =	8 × 5 =	8 × 4 =

Write the missing numbers.

◯ × 8 = 64	9 × ◯ = 54	2 × ◯ = 18	6 × ◯ = 60
5 × ◯ = 35	◯ × 7 = 49	◯ × 7 = 21	◯ × 5 = 45
◯ × 6 = 42	5 × ◯ = 45	6 × ◯ = 30	7 × ◯ = 35
4 × ◯ = 24	◯ × 6 = 36	◯ × 8 = 72	◯ × 9 = 63
◯ × 3 = 27	10 × ◯ = 40	4 × ◯ = 28	8 × ◯ = 56

Which numbers are these machines multiplying by?

3 → × ◯ → ◯ = 21	10 → × ◯ → ◯ = 70	3 → × ◯ → ◯ = 27
7 → × ◯ → ◯ = 42	6 → × ◯ → ◯ = 24	8 → × ◯ → ◯ = 72

Answer these.

What is the fifth multiple of 4? ☐

What is 10 multiplied by 6? ☐

What is eight multiplied by 9? ☐

What is the seventh multiple of 7? ☐

What is double 8? ☐

What is 10 doubled? ☐

What is 6 times seven? ☐

What are seven sixes? ☐

Write the answer chains.

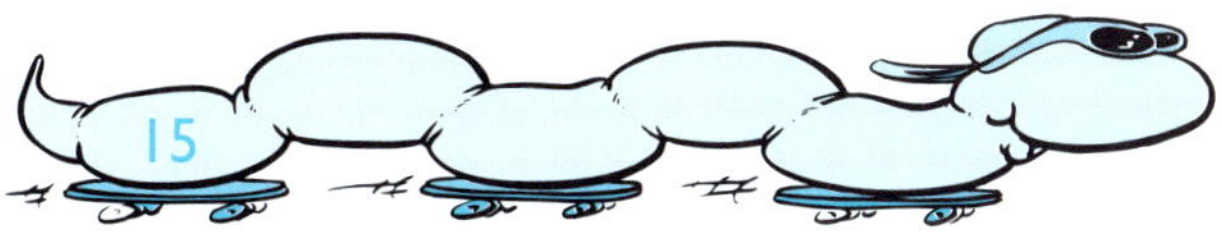

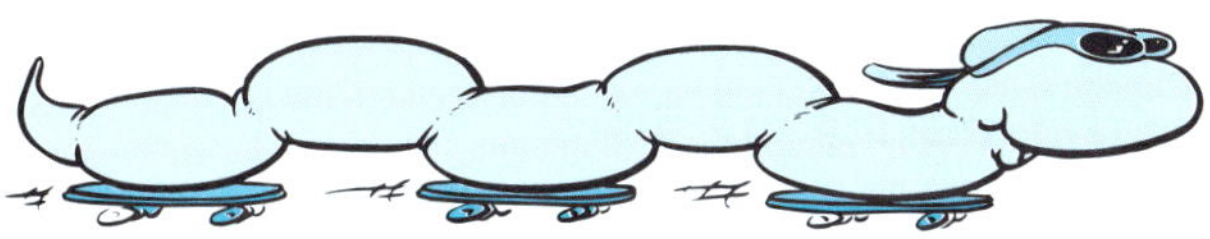

Write different multiplications for these numbers.

☐ × ☐

☐ × ☐

☐ × ☐

☐ × ☐

☐ × ☐

☐ × ☐

☐ × ☐

☐ × ☐

☐ × ☐

☐ × ☐

☐ × ☐

☐ × ☐

☐ × ☐

Answer these.

Division tables

You cannot divide a number by 0.

Dividing by 1 does not change a number.

$$8 \div 1 = 8$$

Answer these as quickly as you can.

$40 \div 4 =$
$30 \div 5 =$
$16 \div 2 =$
$27 \div 3 =$
$18 \div 6 =$

$18 \div 2 =$
$24 \div 4 =$
$72 \div 9 =$
$42 \div 6 =$
$80 \div 8 =$

$63 \div 9 =$
$56 \div 7 =$
$72 \div 8 =$
$36 \div 4 =$
$35 \div 5 =$

$64 \div 8 =$
$36 \div 6 =$
$63 \div 7 =$
$56 \div 8 =$
$81 \div 9 =$

Write the missing numbers.

☐ $\div 3$
☐ $\div 8$
☐ $\div 7$

☐ $\div 6$
☐ $\div 2$
☐ $\div 5$

☐ $\div 9$
☐ $\div 6$
☐ $\div 5$

$24 \div$ ☐
$48 \div$ ☐
$56 \div$ ☐

$30 \div$ ☐
$45 \div$ ☐
$50 \div$ ☐

$18 \div$ ☐
$30 \div$ ☐
$54 \div$ ☐

Write the missing numbers.

$\div 4$

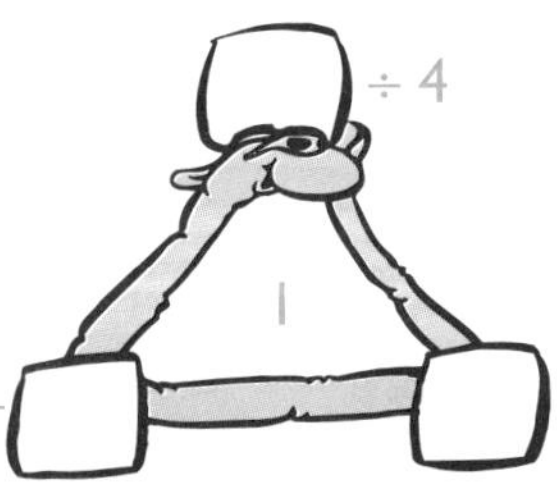

$8 \div$ $\div 7$

$\div 5$

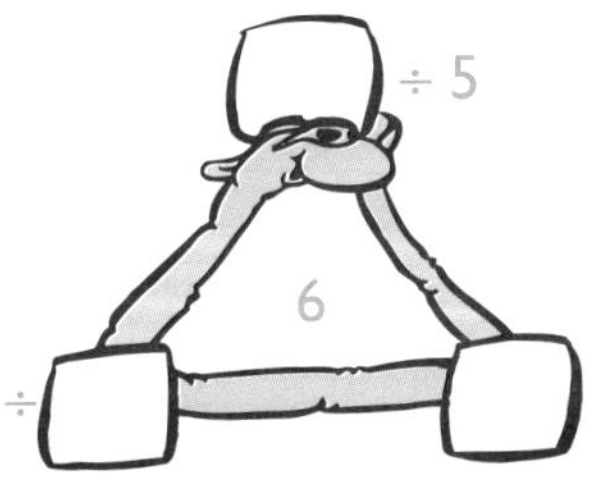

$12 \div$ $\div 7$

$20 \div$

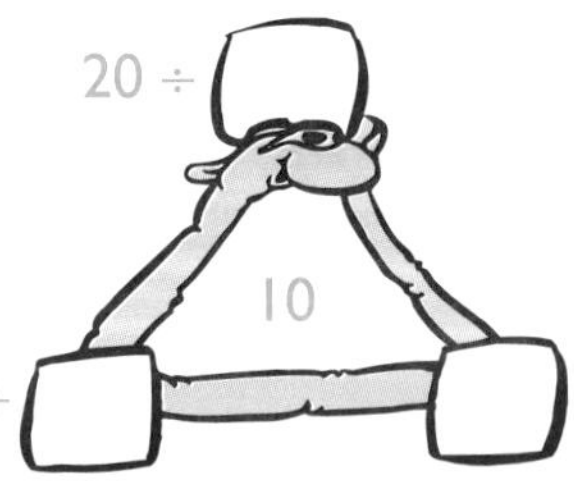

$30 \div$ $\div 10$

Use the code wheel.
Discover the city names.

Answer these divisions.

$\frac{36}{4} = \square$ $\frac{54}{9} = \square$ $\frac{63}{7} - \square$ $\frac{24}{3} = \square$ $\frac{27}{3} = \square$ $\frac{50}{10} = \square$

$\frac{45}{9} = \square$ $\frac{21}{7} = \square$ $\frac{18}{2} = \square$ $\frac{56}{8} = \square$ $\frac{64}{8} = \square$ $\frac{20}{4} = \square$

Write the missing numbers.

16 ÷ ◯ = 2	◯ ÷ 7 = 8	42 ÷ ◯ = 7	◯ ÷ 8 = 3
40 ÷ ◯ = 5	◯ ÷ 3 = 6	54 ÷ ◯ = 9	◯ ÷ 2 = 7
32 ÷ ◯ = 4	◯ ÷ 6 = 4	49 ÷ ◯ = 7	◯ ÷ 5 = 6
12 ÷ ◯ = 3	◯ ÷ 8 = 10	70 ÷ ◯ = 10	◯ ÷ 4 = 8

Multiples and factors

Multiples of 4 are all divisible by 4.

4, 8, 12, 16, 20 → 40

Multiples of 9 are all divisible by 9.

9, 18, 27, 36, 45 → 90

Multiples do not stop at the tenth.

Continue these multiple chains.

Circle the multiples of these numbers.

3	3 10 15 20 27 33
4	24 30 36 45 50 88
6	18 25 36 44 54 66
8	32 40 45 56 81 96
2	18 21 24 33 40 48
5	35 48 50 65 72 80

Write TRUE or FALSE.

- All multiples of 2 are also multiples of 4. ______________
- All multiples of 5 end in 5 or 0. ______________
- All multiples of 3 are also multiples of 5. ______________
- All multiples of 6 are also multiples of 3. ______________
- All multiples of 7 are even. ______________
- All multiples of 3 are also multiples of 6. ______________

Factors divide exactly into numbers.
The factors of 12 are 1, 2, 3, 4, 6 and 12.
1 is a factor of all whole numbers.

List, in order, factors of:

15 ______ 18 ______

21 ______ 24 ______

30 ______ 32 ______

One number in each set is wrong. Cross it out.

factors of 36: 2, 9, 1, 7, 18

factors of 40: 2, 4, 8, 25, 20

factors of 50: 5, 10, 25, 30, 50

factors of 45: 5, 15, 7, 9, 3

factors of 42: 2, 7, 21, 6, 8

factors of 38: 2, 1, 17, 38, 19

What are the missing numbers?

$12 = 2 \times 2 \times$ ___ $8 = 2 \times 2 \times$ ___ $20 = 2 \times 5 \times$ ___

$18 = 3 \times 3 \times$ ___ $24 = 2 \times 2 \times 2 \times$ ___ $30 = 2 \times 3 \times$ ___

$36 = 2 \times 2 \times 3 \times$ ___ $40 = 2 \times 2 \times 2 \times$ ___ $45 = 3 \times 3 \times$ ___

Brackets

You always work out the bracket first.

$$(5 + 2) \times 6$$
$$7 \times 6 = 42$$

Answer these.

$(4 + 6) \times 7 =$
$(10 - 6) \times 5 =$
$(2 \times 3) + 4 =$
$(8 \times 6) - 3 =$

$(3 + 6) \times 8 =$
$(12 - 3) \times 3 =$
$(4 \times 4) + 7 =$
$(5 \times 9) - 7 =$

$(4 + 2) \times 5 =$
$(11 - 6) \times 7 =$
$(5 \times 4) + 8 =$
$(6 \times 8) - 10 =$

$(5 + 5) \times 6 =$
$(10 - 4) \times 9 =$
$(6 \times 6) + 7 =$
$(9 \times 6) - 8 =$

Try these.

$(6 \times 2) + (4 \times 4)$
$12 + 16 = 28$

$(8 \times 8) + (4 \times 5)$

$(7 \times 2) + (7 \times 2)$

$(7 \times 5) + (4 \times 5)$

$(10 \times 6) + (7 \times 8)$

$(9 \times 9) + (2 \times 5)$

$(6 \times 6) - (2 \times 8)$

$(7 \times 8) - (2 \times 8)$

$(8 \times 9) - (3 \times 4)$

$(7 \times 10) - (4 \times 8)$

$(8 \times 5) - (3 \times 3)$

$(4 \times 9) - (6 \times 6)$

Draw the missing brackets.

$4 + 3 \times 2 = 14$
$4 + 3 \times 2 = 10$

$15 - 3 \times 2 = 9$
$15 - 3 \times 2 = 24$

$7 \times 8 - 2 = 54$
$7 \times 8 - 2 = 42$

$9 \times 6 + 4 = 90$
$9 \times 6 + 4 = 58$

$7 + 7 \times 5 = 42$
$7 + 7 \times 5 = 70$

$30 - 5 \times 4 = 10$
$30 - 5 \times 4 = 100$

Answer these.

(4 + 16) ÷ 2 =	(20 + 30) ÷ 5 =	(12 + 12) ÷ 6 =	(40 + 16) ÷ 7 =
(50 – 5) ÷ 9 =	(70 – 6) ÷ 8 =	(60 – 6) ÷ 9 =	(70 – 7) ÷ 7 =
(45 ÷ 5) + 8 =	(36 ÷ 9) + 7 =	(72 ÷ 8) + 10 =	(32 ÷ 4) + 6 =
(50 ÷ 5) – 2 =	(70 ÷ 7) – 7 =	(81 ÷ 9) – 6 =	(64 ÷ 8) – 1 =

Try these.

(15 ÷ 5) + (25 ÷ 5) = ☐	(36 ÷ 9) + (72 ÷ 8) = ☐	(45 ÷ 5) + (63 ÷ 9) = ☐
(36 ÷ 4) + (15 ÷ 5) = ☐	(72 ÷ 9) + (54 ÷ 9) = ☐	(32 ÷ 4) + (24 ÷ 6) = ☐
(81 ÷ 9) – (12 ÷ 4) = ☐	(42 ÷ 7) – (28 ÷ 7) = ☐	(35 ÷ 5) – (8 ÷ 2) = ☐
(60 ÷ 6) – (24 ÷ 8) = ☐	(90 ÷ 10) – (35 ÷ 5)= ☐	(28 ÷ 4) – (7 ÷ 7) = ☐

Which sign +, –, × or ÷ is missing?

(3 ☐ 7) × 8 = 80	5 × (8 ☐ 2) = 30
7 × (9 ☐ 3) = 21	(12 ☐ 4) × 4 = 12
(10 ☐ 4) × 7 = 42	8 × (8 ☐ 8) = 8

(4 ☐ 3) × (10 ☐ 2) = 56	(10 ☐ 2) × (3 ☐ 3) = 45
(3 ☐ 5) – (24 ☐ 2) = 3	(70 ☐ 7) – (42 ☐ 7) = 4
(10 ☐ 7) × (14 ☐ 2) = 21	(3 ☐ 4) – (60 ☐ 10) = 6

Doubles

On a dartboard each outer segment is a "double".

What are these scores?

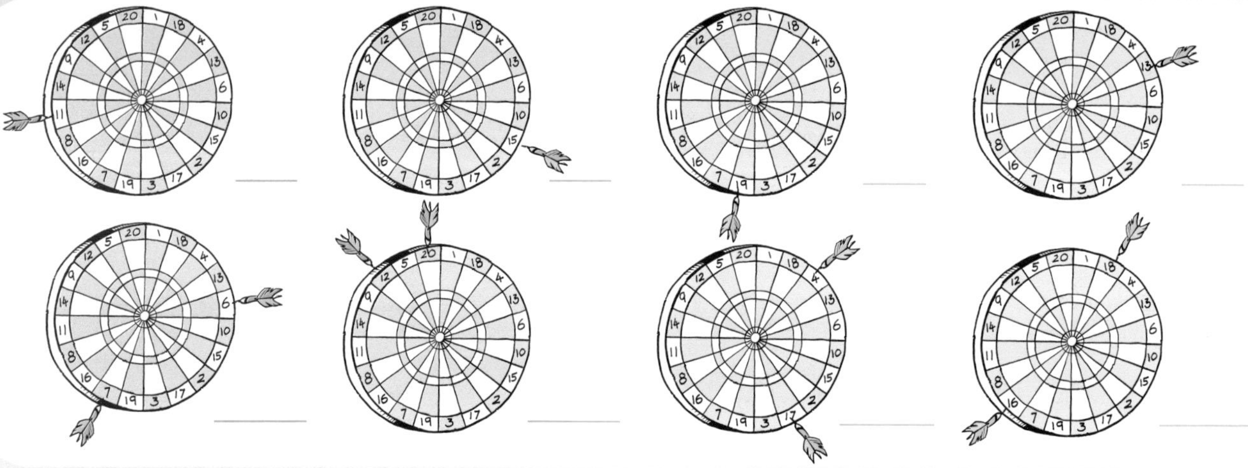

Double each number.

Complete the tables.

IN	46	64	75	79	89	97
OUT						

IN						
OUT	120	144	156	180	196	210

Each container is half full.
Write the full amount.

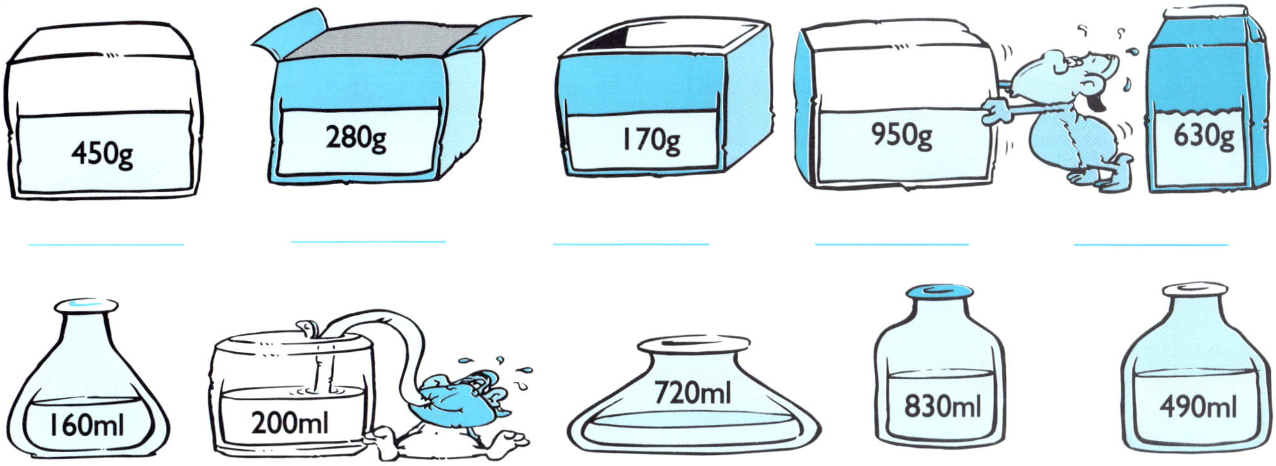

Double these large numbers.

4800	→		1700	→		5900	→		8000	→	
3600	→		2500	→		7700	→		6100	→	

Continue these doubling patterns.

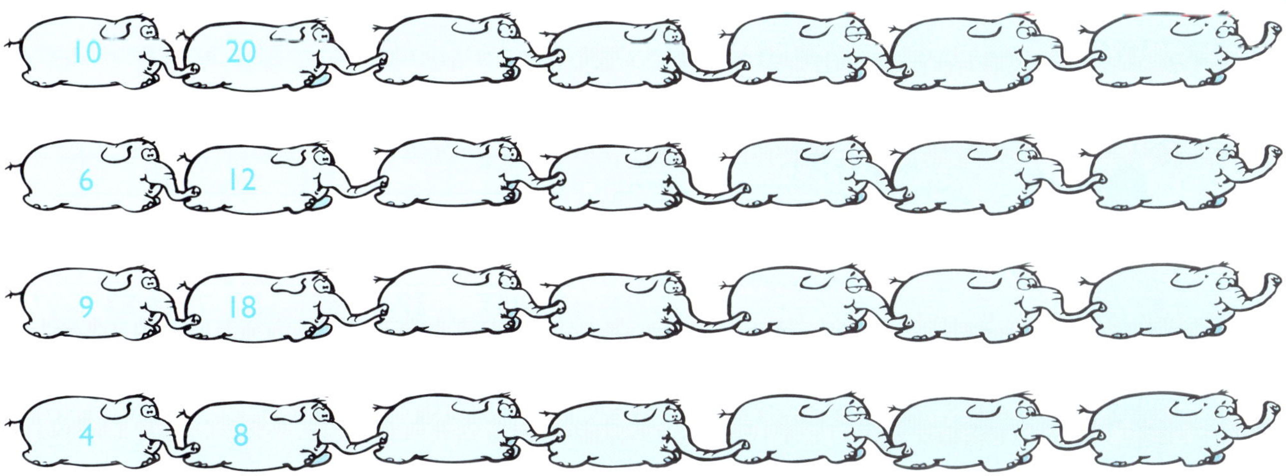

Halving

Which double was thrown to make these scores.

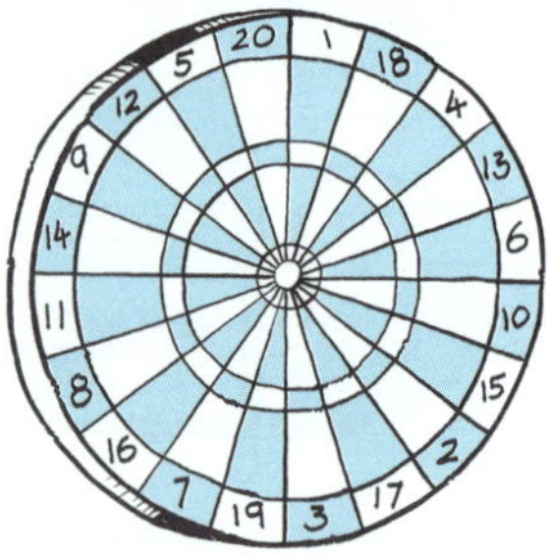

30	24	34	26	40
32	28	22	20	38

Halve each number.

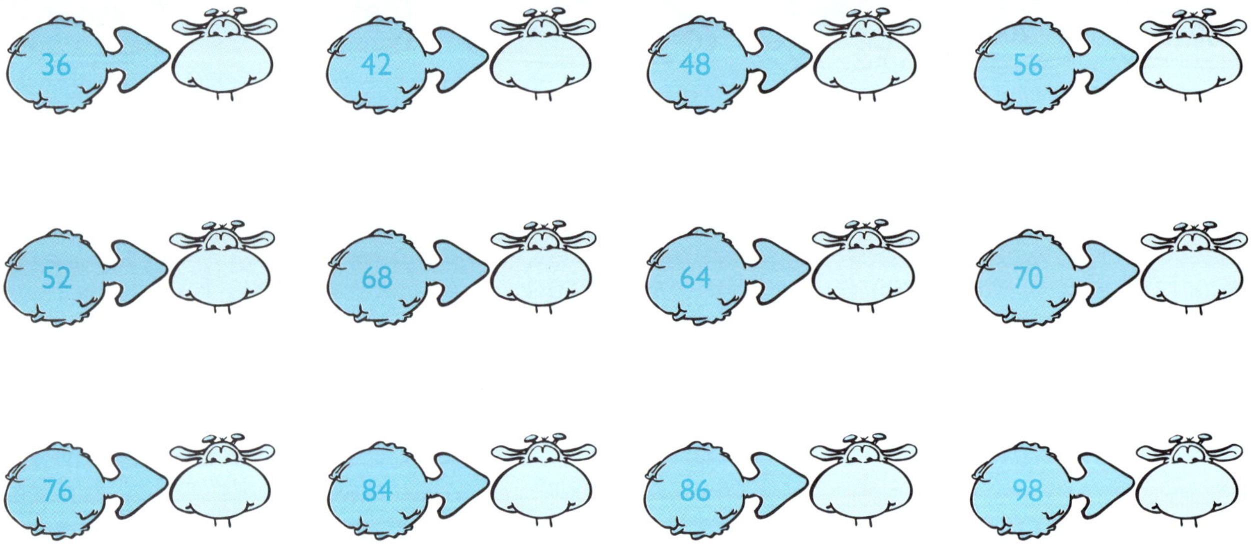

Complete the tables.

IN	158	174	178	188	190	196
OUT						

IN						
OUT	39	45	52	77	83	97

How much will be left after half is poured out?

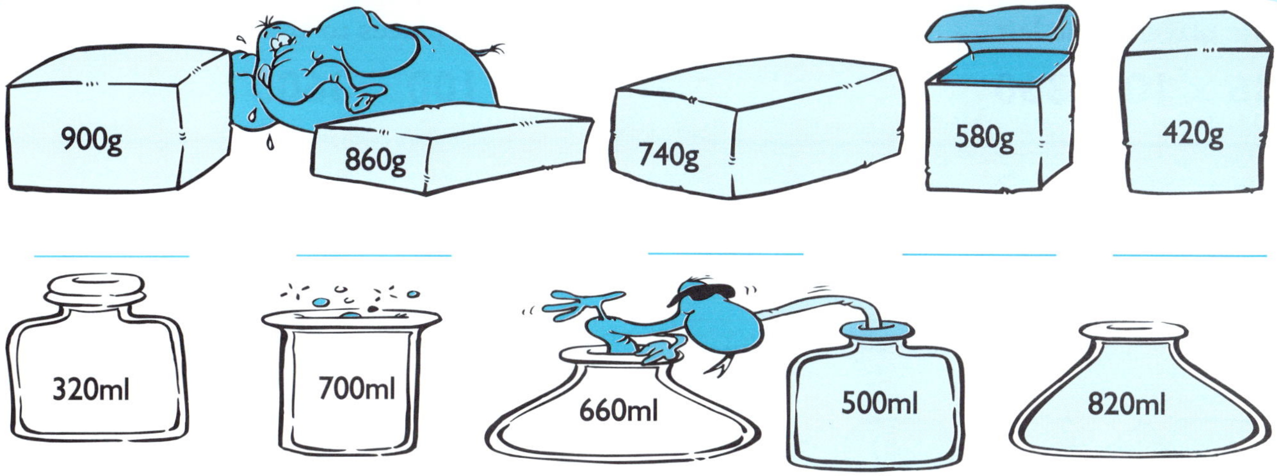

Halve these large numbers.

Continue these halving patterns.

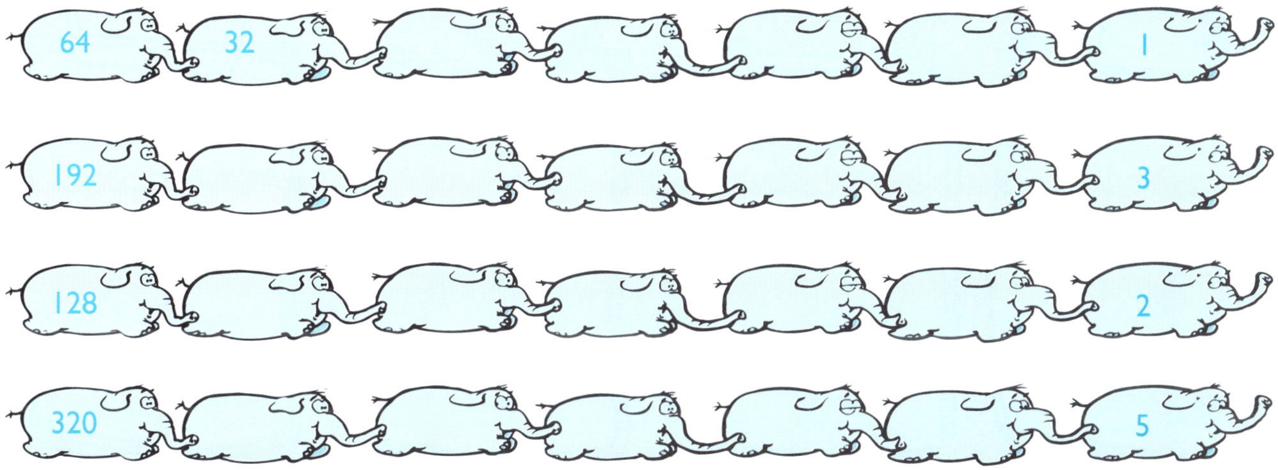

Multiples of 10

When multiplying by 10 the digits jump one place to the left.

36 × 10 = 360

When multiplying by 100 the digits jump two places to the left.

36 × 100 = 3600

Multiply each number by 10.

64	102
70	500
97	640
43	372

Multiply each number by 100.

36	→		400	→	
47	→		320	→	
80	→		704	→	
62	→		638	→	

Multiply each number by the top number.

× 30	× 70	× 200	× 600
40 ______	50 ______	20 ______	30 ______
60 ______	90 ______	40 ______	50 ______
200 ______	300 ______	600 ______	100 ______
400 ______	500 ______	700 ______	400 ______
900 ______	800 ______	900 ______	700 ______

When dividing by 10 the digits jump one place to the right.

5600 ÷ 10 = 560

When dividing by 100 the digits jump two places to the right.

5600 ÷ 100 = 56

Divide each number by 10.

810 →		3940 →	
560 →		7000 →	
300 →		56,000 →	
1020 →		4720 →	

Divide each number by 100.

3400 →		30,700 →	
6900 →		800,000 →	
2000 →		650,000 →	
8000 →		80,000 →	

Divide each number by the top number.

÷ 20

100 ______
460 ______
900 ______
1000 ______
1800 ______

÷ 40

400 ______
1200 ______
16,000 ______
36,000 ______
40,000 ______

÷ 50

1000 ______
2500 ______
40,000 ______
55,000 ______
100,000 ______

÷ 70

700 ______
7000 ______
21,000 ______
350,000 ______
700,000 ______

Multiplying TU by U

When multiplying, break the sum down into simpler parts.

47×6 →

$40 \times 6 = 240$
$7 \times 6 = 42$
Answer 282

Write the answers.

$40 \times 7 =$ $46 \times 7 =$	$50 \times 8 =$ $59 \times 8 =$	$30 \times 9 =$ $37 \times 9 =$	$70 \times 4 =$ $78 \times 4 =$
$50 \times 6 =$ $55 \times 6 =$	$80 \times 3 =$ $87 \times 3 =$	$60 \times 4 =$ $63 \times 4 =$	$80 \times 8 =$ $88 \times 8 =$
$60 \times 9 =$ $62 \times 9 =$	$40 \times 7 =$ $45 \times 7 =$	$90 \times 3 =$ $96 \times 3 =$	$90 \times 8 =$ $99 \times 8 =$

Answer these.

$46 \times 4 =$
$37 \times 5 =$
$25 \times 9 =$
$39 \times 3 =$

$52 \times 4 =$
$49 \times 2 =$
$36 \times 8 =$
$65 \times 9 =$

$76 \times 3 =$
$56 \times 5 =$
$84 \times 7 =$
$83 \times 6 =$

$45 \times 6 =$
$72 \times 8 =$
$62 \times 6 =$
$75 \times 9 =$

Now try these.

42 × 4	56 × 2	78 × 3	55 × 5	72 × 4
44 × 9	67 × 7	95 × 6	83 × 7	

Write the total cost of:

3 @ £45 = £ ________ 6 @ £17 = £ ________ 4 @ £27 = £ ________

5 @ £14 = £ ________ 9 @ £36 = £ ________ 8 @ £32 = £ ________

7 @ £15 = £ ________ 5 @ £76 = £ ________ 6 @ £19 = £ ________

Write the total weight.

8 boxes ______ 7 boxes ______ 5 boxes ______ 4 boxes ______ 6 boxes ______

3 boxes ______ 9 boxes ______ 7 boxes ______ 8 boxes ______ 5 boxes ______

Complete these tables.

×	42	36	25
4			
5			
6			

×	24	32	56
2			
7			
9			

×	17	39	64
3			
5			
10			

Write the missing answers. Circle the answer which is nearest to 500.

Multiplying HTU by U

When multiplying, break the sum down into simpler parts.

236 × 4

200 × 4 = 800
30 × 4 = 120
6 × 4 = 24
Answer 944

Join bats to balls.

Bats: 400 × 3, 800 × 8, 600 × 8, 800 × 4, 700 × 5, 500 × 2, 300 × 9, 900 × 9, 700 × 4

Balls: 4800, 1200, 1000, 8100, 3200, 6400, 2800, 2700, 3500

Answer these.

340 × 2 =	780 × 9 =	660 × 6 =
650 × 4 =	510 × 8 =	740 × 6 =
470 × 3 =	520 × 6 =	850 × 9 =
480 × 4 =	340 × 7 =	530 × 7 =

Write the answers.

Tick the answer nearest to 2000.

Tick the largest answer in each pair.

414 × 5	636 × 3	998 × 8	987 × 6
515 × 4	363 × 6	889 × 9	678 × 9

Make each number seven times bigger.

Write the answers.

$$\begin{array}{r} 434 \\ \times\ 2 \\ \hline \end{array} \qquad \begin{array}{r} 577 \\ \times\ 4 \\ \hline \end{array} \qquad \begin{array}{r} 609 \\ \times\ 5 \\ \hline \end{array}$$

$$\begin{array}{r} 422 \\ \times\ 9 \\ \hline \end{array} \qquad \begin{array}{r} 748 \\ \times\ 6 \\ \hline \end{array} \qquad \begin{array}{r} 627 \\ \times\ 9 \\ \hline \end{array}$$

Write the missing digits.

$$\begin{array}{r} 2\,\square\,9 \\ \times\quad 6 \\ \hline 1674 \end{array} \qquad \begin{array}{r} \square\,0\,9 \\ \times\quad 7 \\ \hline 496\,\square \end{array}$$

$$\begin{array}{r} 5\,6\,\square \\ \times\quad 4 \\ \hline \square\,240 \end{array} \qquad \begin{array}{r} 4\,8\,\square \\ \times\quad 5 \\ \hline \square\,410 \end{array}$$

Multiplying TU by TU

There are many ways of answering 24×30.
Here is one:

	30
20	600
4	120

$$\begin{array}{r} 600 \\ +\ 120 \\ \text{Answer}\ \ 720 \end{array}$$

Join eggs to cups.

30×30 50×70 60×80 40×40 90×60 80×40

4800   900   1600   3200   3500   5400

Answer these.

$47 \times 20 =$
$87 \times 30 =$
$56 \times 80 =$

$57 \times 50 =$
$96 \times 20 =$
$77 \times 90 =$

$59 \times 30 =$
$45 \times 60 =$
$68 \times 80 =$

$64 \times 70 =$
$73 \times 60 =$
$75 \times 40 =$

Answer these.

$\begin{array}{r} 24 \\ \times\ 30 \\ \hline \end{array}$ $\begin{array}{r} 45 \\ \times\ 40 \\ \hline \end{array}$ $\begin{array}{r} 26 \\ \times\ 30 \\ \hline \end{array}$ $\begin{array}{r} 47 \\ \times\ 50 \\ \hline \end{array}$ $\begin{array}{r} 54 \\ \times\ 20 \\ \hline \end{array}$

$\begin{array}{r} 74 \\ \times\ 30 \\ \hline \end{array}$ $\begin{array}{r} 55 \\ \times\ 20 \\ \hline \end{array}$ $\begin{array}{r} 79 \\ \times\ 30 \\ \hline \end{array}$ $\begin{array}{r} 67 \\ \times\ 40 \\ \hline \end{array}$

Here is another way to answer 24 × 30: → 24 × 10 × 3
240 × 3
= 720

Write the total number of pins.

Pins 36	In 40 boxes	☐	Pins 36	In 20 boxes	☐
Pins 24	In 50 boxes	☐	Pins 24	In 30 boxes	☐
Pins 48	In 40 boxes	☐	Pins 48	In 70 boxes	☐
Pins 50	In 24 boxes	☐	Pins 50	In 46 boxes	☐
Pins 70	In 35 boxes	☐	Pins 70	In 64 boxes	☐

Write the missing numbers.

5◯ × 20 = 1060	4◯ × 30 = 1350	1◯ × 50 = 900
2◯ × 70 = 1540	3◯ × 40 = 1280	6◯ × 60 = 3660
5◯ × 50 = 2700	7◯ × 20 = 1480	8◯ × 30 = 2670

Write the missing numbers.

30 × ☐ = 1200	70 × ☐ = 4900	☐ × 90 = 6300
40 × ☐ = 2000	20 × ☐ = 1800	☐ × 60 = 3600
80 × ☐ = 5600	40 × ☐ = 3200	☐ × 70 = 4200

Dividing TU by U

Here is one way to divide.

$$96 \div 4 = 4\overline{)96}$$

$$\begin{array}{rl} 4\overline{)96} & \\ -\ 80 & \rightarrow 20 \times 4 \\ 16 & \\ -\ 16 & \rightarrow 4 \times 4 \end{array}$$

Answer 24

Match kites to posts.

Write the answers.

$4\overline{)84}$ $5\overline{)95}$ $7\overline{)98}$ $2\overline{)76}$ $3\overline{)87}$

$2\overline{)78}$ $3\overline{)69}$ $6\overline{)96}$ $3\overline{)81}$ $4\overline{)76}$

What are the remainders?

$2\overline{)75}$ r ___ $3\overline{)76}$ r ___ $4\overline{)79}$ r ___ $2\overline{)81}$ r ___ $5\overline{)74}$ r ___

$6\overline{)85}$ r ___ $4\overline{)75}$ r ___ $3\overline{)82}$ r ___ $8\overline{)97}$ r ___ $4\overline{)99}$ r ___

Tick the numbers that are not exactly divisible by 3.

86 54 70 74 91 44 69 79

Circle the odd one out.

55 ÷ 6 37 ÷ 3 96 ÷ 5 49 ÷ 4 93 ÷ 4 86 ÷ 6

$\frac{1}{3}$ of each box is used.
How many pins are left?

Pins 96 ______

Pins 81 ______

Pins 75 ______

Pins 66 ______

Pins 87 ______

Write the missing numbers.

☐ ÷ 4
☐ ÷ 3
☐ ÷ 6

☐ ÷ 2
☐ ÷ 5
☐ ÷ 7

☐ ÷ 3
☐ ÷ 4
☐ ÷ 5

Dividing HTU by U

Here is one way to divide.

$8\overline{)344}$

$$
\begin{array}{rl}
8\overline{)344} & \\
-\ 320 & \rightarrow 40 \times 8 \\
24 & \\
-\ 24 & \rightarrow 3 \times 8 \\
\end{array}
$$

Answer 43

Join sums to answers.

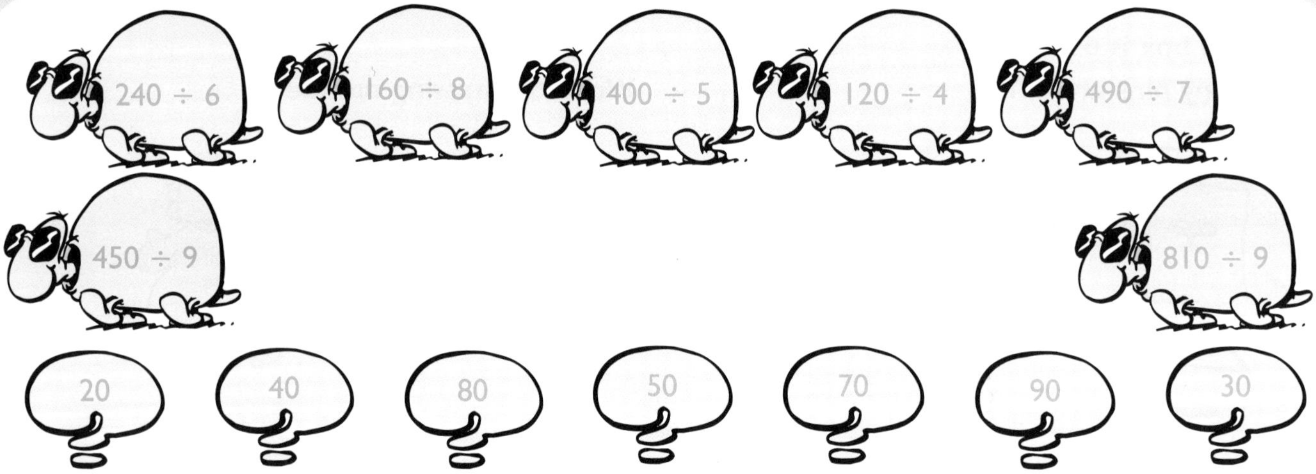

Write the answers.

$2\overline{)768}$ $3\overline{)456}$ $5\overline{)765}$ $4\overline{)716}$ $3\overline{)912}$

$6\overline{)624}$ $7\overline{)847}$ $4\overline{)848}$ $9\overline{)729}$ $8\overline{)584}$

What are the remainders?

$2\overline{)477}$ r ___ $3\overline{)385}$ r ___ $3\overline{)904}$ r ___ $7\overline{)842}$ r ___ $6\overline{)375}$ r ___

$8\overline{)699}$ r ___ $4\overline{)437}$ r ___ $7\overline{)968}$ r ___ $6\overline{)415}$ r ___ $8\overline{)999}$ r ___

Tick the odd one out.

472 ÷ 8	441 ÷ 9	476 ÷ 7	356 ÷ 4
195 ÷ 5	792 ÷ 8	483 ÷ 7	534 ÷ 6

Write the missing digits.

7) □ 3 6 = 4 8	9) 6 1 2 = □ 8	3) 8 4 0 = 2 8 □	5) 8 6 5 = 1 7 □
6) 5 □ 4 = 9 9	4) 7 2 8 = 1 □ 2	8) 9 2 0 = 1 1 □	4) 2 □ 2 = 7 3

Write which of these numbers:

780 116 762

425 677

Is exactly divisible by 5? __________

Is exactly divisible by 6? __________

Has a remainder of 1 when divided by 4? __________

Has a remainder of 2 when divided by 3? __________

$\frac{1}{4}$ of each bottle is left.
How much has been used?

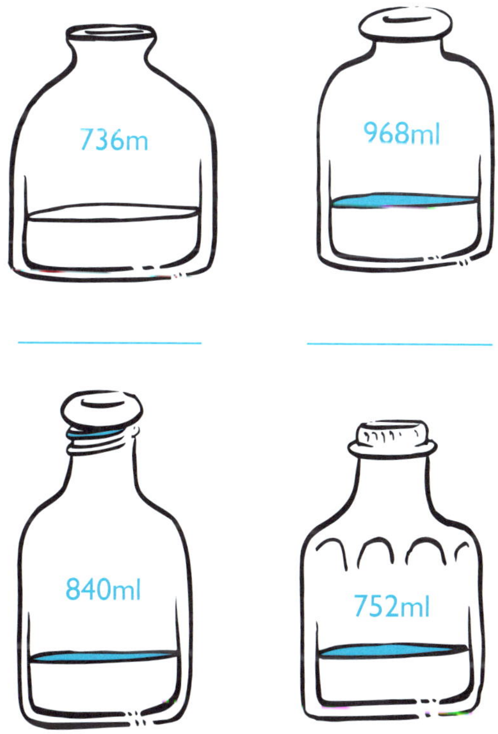

Remainders

Remainders can be written as fractions.

$$\begin{array}{r} 15 \text{ r } 2 \\ 3\overline{)47} \end{array} \quad \text{or} \quad 15\tfrac{2}{3}$$

$$\begin{array}{r} 37 \text{ r } 1 \\ 2\overline{)75} \end{array} \quad \text{or} \quad 37\tfrac{1}{2}$$

Write the remainders as fractions.

$\frac{32}{3} =$	$\frac{25}{3} =$	$\frac{22}{3} =$	$\frac{46}{3} =$	$\frac{56}{3} =$
$\frac{39}{2} =$	$\frac{71}{2} =$	$\frac{87}{2} =$	$\frac{95}{2} =$	$\frac{45}{2} =$
$\frac{36}{5} =$	$\frac{79}{5} =$	$\frac{84}{5} =$	$\frac{47}{5} =$	$\frac{68}{5} =$

Write the missing numbers.

$13\frac{1}{2} = 2\overline{)\square}$	$16\frac{1}{2} = 2\overline{)\square}$	$27\frac{1}{2} = 2\overline{)\square}$	$56\frac{1}{2} = 2\overline{)\square}$
$14\frac{2}{3} = 3\overline{)\square}$	$22\frac{2}{3} = 3\overline{)\square}$	$33\frac{2}{3} = 3\overline{)\square}$	$54\frac{1}{3} = 3\overline{)\square}$
$12\frac{3}{8} = 8\overline{)\square}$	$26\frac{5}{8} = 8\overline{)\square}$	$30\frac{7}{8} = 8\overline{)\square}$	$34\frac{1}{8} = 8\overline{)\square}$

Circle the divisions which will have no remainder.

$2\overline{)225}$	$5\overline{)225}$	$8\overline{)225}$	$6\overline{)225}$	$3\overline{)225}$
$2\overline{)416}$	$5\overline{)416}$	$8\overline{)416}$	$6\overline{)416}$	$3\overline{)416}$

Sometimes remainders tell you to round up, sometimes to round down.

Egg boxes hold 6 eggs.
How many egg boxes are needed for:

58 eggs	79 eggs	87 eggs	99 eggs	100 eggs
127 eggs	175 eggs	204 eggs	258 eggs	347 eggs

Look at the ticket prices.
How many tickets can be bought with:

Ticket £8	Ticket £5	Ticket £3	Ticket £4	Ticket £9
£97	£123	£87	£99	£127
______	______	______	______	______
Ticket £2	Ticket £6	Ticket £7	Ticket £10	Ticket £4
£135	£175	£200	£145	£199
______	______	______	______	______

Money problems

$$£1.25 \times 6 = \begin{array}{r} £1.25 \\ \times\ 6 \\ \hline £7.50 \end{array}$$

$$£4.50 \times 7 = \begin{array}{r} £4.50 \\ \times\ 7 \\ \hline £31.50 \end{array}$$

What is the total cost of these parcels?

cost of 6	cost of 7	cost of 8	cost of 9	cost of 6	cost of 8
£ ______	£ ______	£ ______	£ ______	£ ______	£ ______

Total these bills.

6 pairs of socks @ £3.99
2 pairs of socks @ £4.25 ______
TOTAL £

4 ties @ £7.75
2 shirts @ £9.99 ______
TOTAL £

8 T-shirts @ £7.75
6 pairs of shorts @ £9.99 ______
TOTAL £

3 vests @ £4.65
4 pairs of pants @ £2.35 ______
TOTAL £

Complete these questions.

- Lee has £7.46. Tommy has twice as much.
  How much have they altogether? £ ______
- Wendy has twice as much as Gregory has. Altogether they have £9.90.
  How much does Gregory have? £ ______
- Ali has three times as much as Helen has. Helen has £4.99.
  How much do they have altogether? £ ______

$$£3 \div 2 = 2\overline{)3.00} = £1.50 \quad (\text{quotient } 1.50)$$

$$£3 \div 4 = 4\overline{)3.00} = £0.75 = 75p \quad (\text{quotient } 0.75)$$

Write how much one of each of these would cost.

2 for £3 1 costs £ ______	10 for £7 1 costs £ ______	5 for £7 1 costs £ ______
4 for £10 1 costs £ ______	5 for £12 1 costs £ ______	8 for £18 1 costs £ ______

What is $\frac{1}{4}$ of these prices?

£4.60 £8.24 £7.44 £7.96 £3.40

What is $\frac{1}{10}$ of these prices?

£15 £28

£19 £36

How many coins in each money box?

2ps £7.48 ☐

5ps £6.55 ☐

2ps £8.62 ☐

10ps £3.60 ☐

Using multiplication and division

$24 \times 13 = 312$

$13 \times 24 = 312$

$312 \div 24 = 13$

$312 \div 13 = 24$

Use the first fact to work out the second.

$18 \times 34 = 612$	$47 \times 15 = 705$	$23 \times 19 = 437$
$612 \div 34 =$ ☐	$705 \div 15 =$ ☐	$19 \times 23 =$ ☐
$37 \times 16 = 592$	$29 \times 13 = 377$	$26 \times 22 = 572$
$37 \times 17 =$ ☐	$29 \times 14 =$ ☐	$26 \times 23 =$ ☐
$24 \times 2 = 48$	$36 \times 3 = 108$	$54 \times 6 = 324$
$24 \times 4 =$ ☐	$36 \times 6 =$ ☐	$54 \times 12 =$ ☐

Use answers from the first set of tables to help you answer the second.

$1 \times 8 =$	$1 \times 16 =$	$1 \times 12 =$	$1 \times 24 =$
$2 \times 8 =$	$2 \times 16 =$	$2 \times 12 =$	$2 \times 24 =$
$3 \times 8 =$	$3 \times 16 =$	$3 \times 12 =$	$3 \times 24 =$
$4 \times 8 =$	$4 \times 16 =$	$4 \times 12 =$	$4 \times 24 =$
$5 \times 8 =$	$5 \times 16 =$	$5 \times 12 =$	$5 \times 24 =$
$6 \times 8 =$	$6 \times 16 =$	$6 \times 12 =$	$6 \times 24 =$
$7 \times 8 =$	$7 \times 16 =$	$7 \times 12 =$	$7 \times 24 =$
$8 \times 8 =$	$8 \times 16 =$	$8 \times 12 =$	$8 \times 24 =$
$9 \times 8 =$	$9 \times 16 =$	$9 \times 12 =$	$9 \times 24 =$
$10 \times 8 =$	$10 \times 16 =$	$10 \times 12 =$	$10 \times 24 =$

Page 2

20	63	16	16
42	25	63	24
45	42	36	54
48	64	60	49
56	30	40	32

8	6	9	10
7	7	3	9
7	9	5	5
6	6	9	7
9	4	7	7

7	7	9
6	4	9

Page 3

20	16
60	20
72	42
49	42

15, 42, 64, 40, 48, 16; 60, 72, 24, 36, 24, 63

$21 = 3 \times 7, 7 \times 3$
$24 = 6 \times 4, 4 \times 6, 12 \times 2, 2 \times 12, 3 \times 8, 8 \times 3$
$36 = 6 \times 6, 4 \times 9, 9 \times 4, 3 \times 12, 12 \times 3$
$30 = 10 \times 3, 3 \times 10, 5 \times 6, 6 \times 5$

$9 \times 5 = 45, 5 \times 6 = 30, 10 \times 8 = 80, 9 \times 4 = 36$

Page 4

10	9	7	8
6	6	8	6
8	8	9	9
9	7	9	7
3	10	7	9

$12 \div 3 = 4, 32 \div 8 = 4, 28 \div 7 = 4$
$24 \div 3 = 8, 48 \div 6 = 8, 56 \div 7 = 8$
$42 \div 6 = 7, 14 \div 2 = 7, 35 \div 5 = 7$
$30 \div 6 = 5, 45 \div 9 = 5, 50 \div 10 = 5$
$27 \div 9 = 3, 18 \div 6 = 3, 15 \div 5 = 3$
$18 \div 3 = 6, 30 \div 5 = 6, 54 \div 9 = 6$
1: 4, 8, 7; 6: 30, 2, 42; 10: 2, 3, 100

Page 5

ROME, MADRID, LEEDS, PARIS
9, 6, 9, 8, 9, 5
5, 3, 9, 7, 8, 5

8	56	6	24
8	18	6	14
8	24	7	30
4	80	7	32

Page 6

40, 45, 50, 55, 60, 65
48, 54, 60, 66, 72, 78
32, 36, 40, 44, 48, 52
72, 81, 90, 99, 108, 117
Multiples of 3: 3, 15, 27, 33
Multiples of 6: 18, 36, 54, 66
Multiples of 2: 18, 24, 40, 48
Multiples of 4: 24, 36, 88
Multiples of 8: 32, 40, 56, 96
Multiples of 5: 35, 50, 65, 80

False	True
False	True
False	False

Page 7

Factors of 15: 1, 3, 5, 15
Factors of 21: 1, 3, 7, 21
Factors of 30: 1, 2, 3, 5, 6, 10, 15, 30
Factors of 18: 1, 2, 3, 6, 9, 18
Factors of 24: 1, 2, 3, 4, 6, 8, 12, 24
Factors of 32: 1, 2, 4, 8, 16, 32
Factors of 36: 2, 9, 1, 18 - cross out 7
Factors of 40: 2, 4, 8, 20 - cross out 25
Factors of 50: 5, 10, 25, 50 - cross out 30
Factors of 45: 5, 15, 9, 3 - cross out 7
Factors of 42: 2, 7, 21, 6 - cross out 8
Factors of 38: 2, 1, 38, 19 - cross out 17

3	2	2
2	3	5
3	5	5

Page 8

70	72	30	60
20	27	35	54
10	23	28	43
45	38	38	46

$12 + 16 = 28$ $64 + 20 = 84$ $14 + 14 = 28$
$35 + 20 = 55$ $60 + 56 = 116$ $81 + 10 = 91$
$36 - 16 = 20$ $56 - 16 = 40$ $72 - 12 = 60$
$70 - 32 = 38$ $40 - 9 = 31$ $36 - 36 = 0$

$(4 + 3) \times 2 = 14$ $4 + (3 \times 2) = 10$
$9 \times (6 + 4) = 90$ $(9 \times 6) + 4 = 58$
$15 - (3 \times 2) = 9$ $(15 - 3) \times 2 = 24$
$7 + (7 \times 5) = 42$ $(7 + 7) \times 5 = 70$
$(7 \times 8) - 2 = 51$ $7 \times (8 - 2) = 42$
$30 - (5 \times 4) = 10$ $(30 - 5) \times 4 = 100$

Page 9

10	10	4	8
5	8	6	9
17	11	19	14
8	3	3	7

8	13	16
12	14	12
6	2	3
7	2	6

$(3 + 7) \times 8 = 80$ $5 \times (8 - 2) = 30$
$7 \times (9 \div 3) = 21$ $(12 \div 4) \times 4 = 12$
$(10 - 4) \times 7 = 42$ $8 \times (8 \div 8) = 8$

$(4 + 3) \times (10 - 2) = 56$
$(3 \times 5) - (24 \div 2) = 3$
$(10 - 7) \times (14 \div 2) = 21$
$(10 \div 2) \times (3 \times 3) = 45$
$(70 \div 7) - (42 \div 7) = 4$
$(3 \times 4) - (60 \div 10) = 6$

Page 10

22	30	38	26
26	64	42	68

doubles: 32, 64; 55, 110; 85, 170; 45, 90; 68, 136; 88, 176; 38, 76; 59, 118; 93, 186; 47, 94; 76, 152; 99, 198

IN	46	64	75	79	89	97
OUT	92	128	150	158	178	194

IN	60	72	78	90	98	105
OUT	120	144	156	180	196	210

Page 11

900g, 560g, 340g, 1900g, 1260g,
320ml, 400ml, 1440ml, 1660ml, 980ml
Doubles: 4800, 9600; 1700, 3400; 5900, 11,800; 8000, 16,000; 3600, 7200; 2500, 5000; 7700, 15,400; 6100, 12,200
10, 20, 40, 80, 160, 320, 640
6, 12, 24, 48, 96, 192, 384
9, 18, 36, 72, 144, 288, 576
4, 8, 16, 32, 64, 128, 256

Page 12

Doubles; 30, 15; 24, 12; 34, 17; 26, 13; 40, 20; 32, 16; 28, 14; 22, 11; 20, 10; 38, 19

18	21	24	28
26	34	32	35
38	42	43	49

OUT	79	87	89	94	95	98

IN	78	90	104	154	166	194

Page 13

Halves: 900g, 450g; 860g, 430g; 740g, 370g; 580g, 290g; 420g, 210g
320ml, 160ml; 700ml, 350ml; 660ml, 330ml; 500ml, 250ml; 820ml, 410ml

3500	4500	2500	5000
2300	2900	3100	3700

64, 32, 16, 8, 4, 2, 1
192, 96, 48, 24, 12, 6, 3
128, 64, 32, 16, 8, 4, 2
320, 160, 80, 40, 20, 10, 5

Page 14

Multiply by 10: 640, 1020,
700, 5000,
970, 6400,
430, 3720
Multiply by 100: 3600, 40,000,
4700, 32,000,
8000, 70,400,
6200, 63,800
$\times$ 30: 1200, 1800, 6000, 12,000, 27,000
$\times$ 70: 3500, 6300, 21,000, 35,000, 56,000
$\times$ 200: 4000, 8000, 120,000, 140,000, 180,000
$\times$ 600: 18,000, 30,000, 60,000, 240,000, 420,000

Page 15

Divide by 10: 81, 394,
56, 700,
30, 5600,
102, 472
Divide by 100: 34, 307,

Answers

69, 8000,
20, 6500,
80, 800,
÷ 20: 5, 23, 45, 50, 90
÷ 40: 10, 30, 400, 900, 1000
÷ 50: 20, 50, 800, 1100, 2000
÷ 70: 10, 100, 300, 5000, 10,000

Page 16

$40 \times 7 = 280$ $50 \times 8 = 400$
$46 \times 7 = 322$ $59 \times 8 = 472$

$50 \times 6 = 300$ $80 \times 3 = 240$
$55 \times 6 = 330$ $87 \times 3 = 261$

$60 \times 9 = 540$ $40 \times 7 = 280$
$62 \times 9 = 558$ $45 \times 7 = 315$

$30 \times 9 = 270$ $70 \times 4 = 280$
$37 \times 9 = 333$ $78 \times 4 = 312$

$60 \times 4 = 240$ $80 \times 8 = 640$
$63 \times 4 = 252$ $88 \times 8 = 704$

$90 \times 3 = 270$ $90 \times 8 = 720$
$96 \times 3 = 288$ $99 \times 8 = 792$

184	208	228	270
185	98	280	576
225	288	588	372
117	585	498	675

$42 \times 4 = 168$, $56 \times 2 = 112$, $78 \times 3 = 234$,
$55 \times 5 = 275$, $72 \times 4 = 288$
$44 \times 9 = 396$, $67 \times 7 = 469$, $95 \times 6 = 570$,
$83 \times 7 = 581$

Page 17

3 @ £45 = £135, 5 @ £14 = £70,
7 @ £15 = £105, 6 @ £17 = £102,
9 @ £36 = £324, 5 @ £76 = £380,
4 @ £27 = £108, 8 @ £32 = £256,
6 @ £19 = £114
8 boxes @ 96g = 768g, 7 boxes @ 48g = 336g,
5 boxes @ 75g = 375g, 4 boxes @ 63g = 252g,
6 boxes @ 74g = 444g, 3 boxes @ 37g = 111g,
9 boxes @ 55g = 495g, , 7 boxes @ 82g = 574g,
8 boxes @ 42g = 336g, 5 boxes @ 94g = 470g
×4: 168, 144, 100 ×2: 48, 64, 112
×5: 210, 180, 125 ×7: 168, 224, 392
×6: 252, 216, 150 ×9: 216, 288, 504

×3: 51, 117, 192
×5: 85, 195, 320
×10: 170, 390, 640
512, 423, 468, 435, 380, (511,) 483, 432

Page 18

$400 \times 3 = 1200$, $600 \times 8 = 4800$, $700 \times 5 = 3500$,
$300 \times 9 = 2700$, $900 \times 9 = 8100$, $800 \times 8 = 6400$,
$800 \times 4 = 3200$, $500 \times 2 = 1000$, $700 \times 4 = 2800$

680	7020	3960
2600	4080	4440
1410	3120	7650
1920	2380	3710

Page 19

1784, 2637, 2037, 2187, (1992,) 2445, 2512

largest: 414×5, 363×6, 889×9, 678×9
3262, 5047, 3052, 3794, 2646, 6468
$434 \times 2 = 868$, $577 \times 4 = 2308$, $609 \times 5 = 3045$
$422 \times 9 = 3798$, $748 \times 6 = 4488$, $627 \times 9 = 5643$
$279 \times 6 = 1674$, $709 \times 7 = 4963$, $560 \times 4 = 2240$
$482 \times 5 = 2410$

Page 20

$30 \times 30 = 900$, $50 \times 70 = 3500$, $60 \times 80 = 4800$,
$40 \times 40 = 1600$, $90 \times 60 = 5400$, $80 \times 40 = 3200$

940	2850	1770	4480
2610	1920	2700	4380
4480	6930	5440	3000

720, 1800, 780, 2350, 1080
2220, 1100, 2370, 2680

Page 21

1440	720
1200	720
1920	3360
1200	2300
2450	4480

53	45	18
22	32	61
54	74	89

40	70	70
50	90	60
70	80	60

Page 22

$72 \div 3 = 24$, $91 \div 7 = 13$, $96 \div 2 = 48$,
$84 \div 6 = 14$, $95 \div 5 = 19$, $68 \div 4 = 17$,
$75 \div 3 = 25$
21, 19, 14, 38, 29
39, 23, 16, 27, 19
37 r.1, 25 r.1, 19 r.3, 40 r.1, 14 r.4
14 r.1, 18 r.3, 27 r.1, 12 r.1, 24 r.3

Page 23

Numbers not divisible by 3: 86, 70, 74, 91, 44, 79
Odd one out: 86 ÷ 6 (as the answer has a remainder of 2, while the others all have remainder of 1)
64, 54, 50, 44, 58
56, 42, 84
26, 65, 91
45, 60, 75

Page 24

$450 \div 9 = 50$, $240 \div 6 = 40$, $160 \div 8 = 20$,
$400 \div 5 = 80$, $120 \div 4 = 30$, $490 \div 7 = 70$,
$810 \div 9 = 90$
384, 152, 153, 179, 304
104, 121, 212, 81, 73

238 r.1, 128 r.1, 301 r.1, 120 r.2, 62 r.3
87 r.3, 109 r.1, 138 r.2, 69 r.1, 124 r.7

Page 25

The odd one out: 476 ÷ 7 (as it has the only answer of the set which doesn't end in '9')
3 6 0 3
9 8 5 9
780, 425; 780, 762; 425, 677; 116, 425, 677
552ml, 726ml, 630ml, 564ml

Page 26

$10\frac{2}{3}$, $8\frac{1}{3}$, $7\frac{1}{3}$, $15\frac{1}{3}$, $18\frac{2}{3}$

$19\frac{1}{2}$, $35\frac{1}{2}$, $43\frac{1}{2}$, $47\frac{1}{2}$, $22\frac{1}{2}$

$7\frac{1}{5}$, $15\frac{4}{5}$, $16\frac{4}{5}$, $9\frac{2}{5}$, $13\frac{3}{5}$

27	33	55	113
44	68	101	163
99	213	247	273

divisions which have no remainders:
$2\overline{)416}$, $5\overline{)225}$ $8\overline{)416}$ $3\overline{)225}$

Page 27

Boxes needed: 10, 14, 15, 17, 17
Boxes needed: 22, 30, 34, 43, 58
12, 24, 29, 24, 14
67, 29, 28, 14, 49

Page 28

£9.00, £15.75, £24.40, £40.50, £34.50, £39.92

£23.94	£31.00
£ 8.50	£19.98
£32.44	£50.98

£ 62.00	£13.95
£ 59.94	£ 9.40
£121.94	£23.35

£22.38, £3.30, £19.96

Page 29

£1.50, £0.70, £1.40
£2.50, £2.40, £2.25
£1.15, £2.06, £1.86, £1.99, £0.85
£15 = £1.50, £28 = £2.80,
£19 = £1.90, £36 = £3.60
374, 131, 431, 36

Page 30

18	47	437
629	406	598
96	216	648

8, 16, 24, 32, 40, 48, 56, 64, 72, 80
16, 32, 48, 64, 80, 96, 112, 128, 144, 160
12, 24, 36, 48, 60, 72, 84, 96, 108, 120
24, 48, 72, 96, 120, 144, 168, 192, 216, 240